More or fewer?

T0173581

Objective

I have developed a sense of comparing collections.

1)

2)

3)

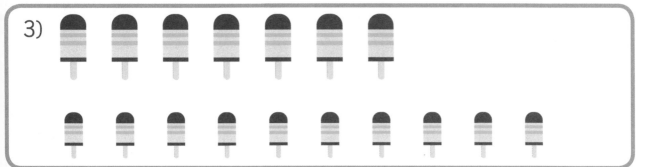

Teacher's notes

1) Tick which collection has the most in it.

2) Are there more bananas or apples?

3) Circle which collection has 'fewer than' the other collection?

3

Colour match

Objective

I can find the size of a collection.

You need

Coloured pencils

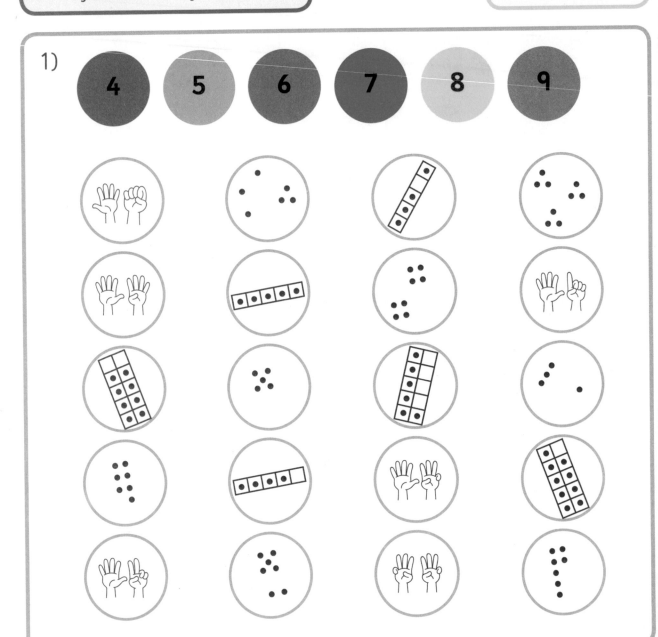

1)

Teacher's notes

1) Colour the dot or finger pattern to match with the correct number.

4

Early Level Maths

Record Book

Aligned to CfE benchmarks

Contents

Regular patterns

Objective

I can estimate and check size.

1)

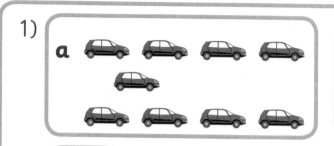

a | b

⬚ has more.   **a** has ⬚   and **b** has ⬚

2)

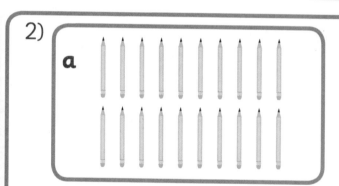
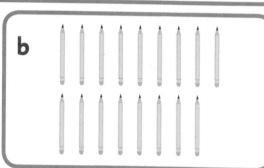

a | b

⬚ has the fewest.   **a** has ⬚   and **b** has ⬚

3)

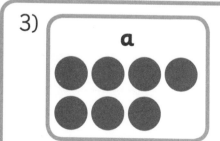

a | b

⬚ has more.

a has ⬚   and **b** has ⬚

Teacher's notes

1) Which has more?
   a or b? Check by
   how many.

2) Who has the fewest?
   a or b? Check by
   how many.

3) Who has the most?
   a or b? Check by
   how many.

Objective

I can count in order forwards and backwards.

1)

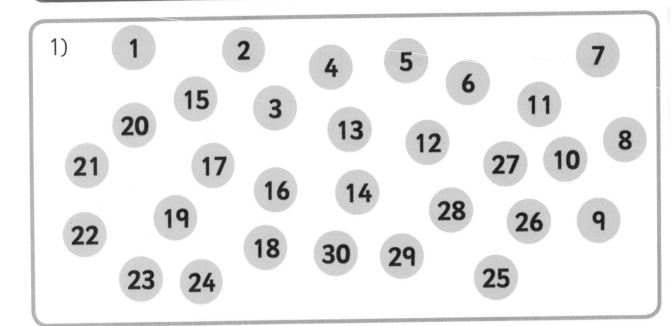

2)

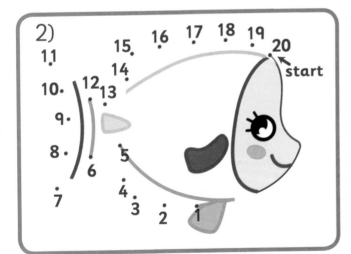

start

3)

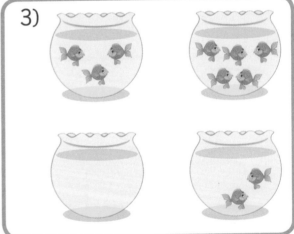

Teacher's notes

1) Start at 1 and draw a line to the next number in the forward word sequence up to 30.

2) Start at 20, fill in the dot-to-dot picture following the backward word sequence from 20 to 1.

3) Circle any fish bowls with zero fish.

Me first!

Objective

I can put numbers and objects in order.

1)

Before	In between	After
◯ ← 6	13 ◯ 15	6 → ◯
◯ ← 12	8 ◯ 10	29 → ◯
◯ ← 8	18 ◯ 20	14 → ◯
◯ ← 20	11 ◯ 13	8 → ◯

2)

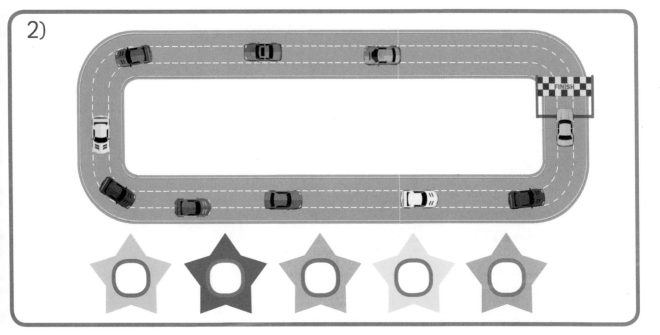

Teacher's notes

1) Record the missing numbers in the boxes.
2) Record the position that each of the colours finished in the race.

Date completed

Objective

I can put numbers in order.

Teacher's notes

1) Fill in the numbers from the forwards and backwards sequence to 20 in the skyscrapers.

2) Write the numbers in order, starting with the smallest number.

3) Write the numbers in order, starting with the largest number.

1)
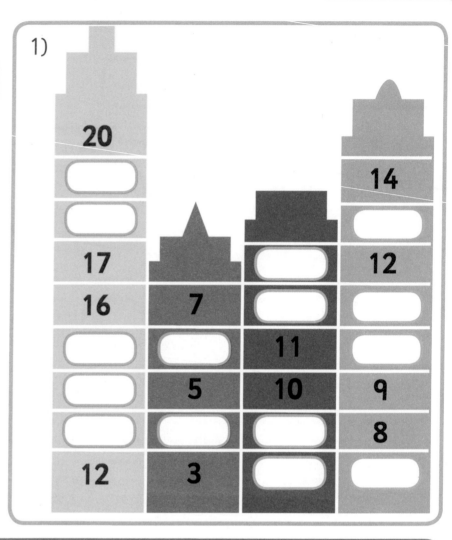

2)

3
9   18
   15
3   11

3)

19
10   19
   16
5   11

Collections

Objective

I can count and compare collections.

1)

a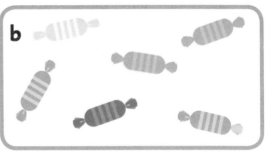

b

Yes ⬜

No ⬜

2)

a

b

Yes ⬜

No ⬜

3)

a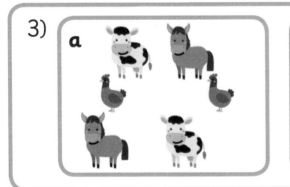

b

Yes ⬜

No ⬜

4)

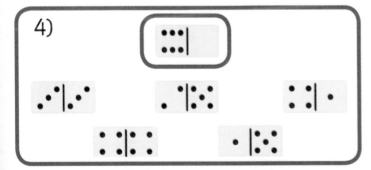

Teacher's notes

1) Are these two collections the same size?
2) Are these two collections the same size?
3) Are these two collections the same size?
4) Draw a line from this first domino to any other pattern of 6.

9

Date completed

Objective

I can count and record how many.

1)

2)

| 1 | 2 | 3 | 4 | 5 | 6 | 7 | 8 | 9 | 10 | 11 | 12 | 13 | 14 | 15 | 16 | 17 | 18 | 19 | 20 |

| 1 | 2 | 3 | 4 | 5 | 6 | 7 | 8 | 9 | 10 | 11 | 12 | 13 | 14 | 15 | 16 | 17 | 18 | 19 | 20 |

Teacher's notes

1) Count the flowers and record in the box.

2) Count the counters and record in the box.

Party time

Objective

I can find the total of a collection.

1)

12   18   11   14   16   15   19   17   20

2)

I have 7 balloons.

I have 16 balloons.

I have 15 balloons.

Teacher's notes

1) Circle the correct number of sweets.

2) Tick if the child has counted the number of balloons correctly.

11

Colour the facts

Objective

I can partition quantities into parts.

You need

Coloured pencils

1)

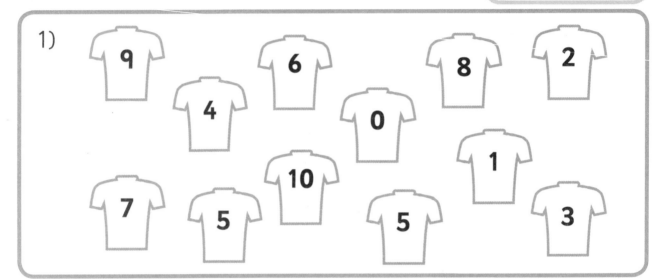

2)

Teacher's notes

1) Colour the pairs that make ten the same colour.

2) Using two or three colours, colour in the strips to show the number facts to ten. Record as a sum in the box.

Finding totals

Objective

I can use mathematical symbols.
I can count on and back to add and subtract.

1)

2)

3)

$3 + 4 =$ ☐ $7 - 2 =$ ☐

$5 + 3 =$ ☐ $9 - 6 =$ ☐

Teacher's notes

1) Look at the picture and write a calculation to show how many apples there are altogether.

2) Look at the picture. How many acorns are left after the squirrel has taken 1? You can show your calculation in the box.

3) Find the answers using the number line.

Date completed

Objective

I can solve simple missing number problems.

You need

Coloured pencils

1)

10	
3	

8	
6	

9	
5	

5	
2	

6	
3	

7	
	1

2)

$7 + \boxed{} = 10$

Teacher's notes

1) Write in the missing numbers to complete the bar diagrams.

2) Draw red and green apples into the ten frame to complete the missing number calculation – **There are ten apples. 7 are green. How many are red?**

Double up

Objective

I can double numbers to a total of ten mentally.

1)

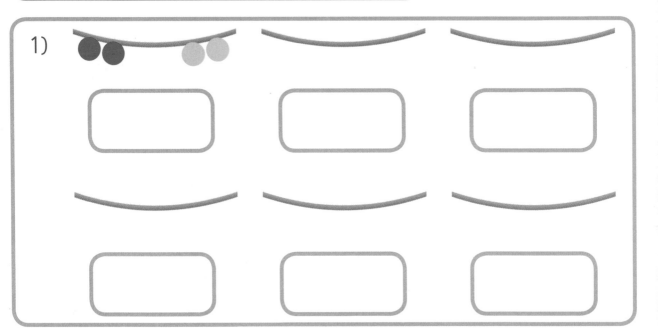

2)

Double 5

Double 0

3 plus 3

Two groups of 4

1 add 1

Double 2

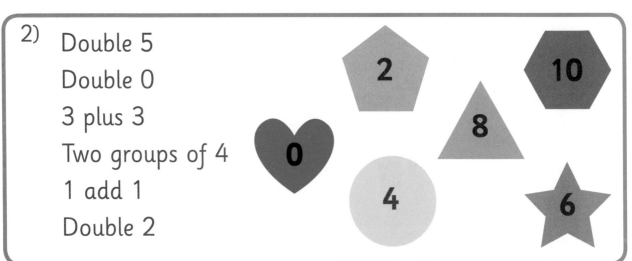

Teacher's notes

1) Show me all the doubles facts you know to 10 by drawing red and green circles on each string. Write the double fact for each string in the box.

2) Match each question to the correct answer.

Objective

I can add and subtract to 10.

1)

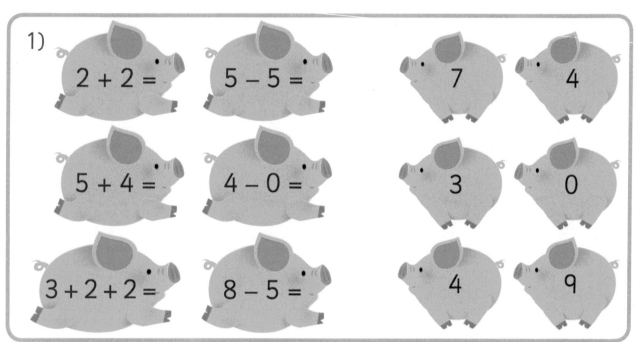

2 + 2 =   5 – 5 =   7   4

5 + 4 =   4 – 0 =   3   0

3 + 2 + 2 =   8 – 5 =   4   9

2)

Teacher's notes

1) Match each question to its correct answer.

2) Four cows were in a field. Two more cows joined them. How many cows are in the field now?

The bigger half?

Objective

I can split a whole into smaller parts and explain that equal parts are the same size.

I can use vocabulary to describe halves.

1)

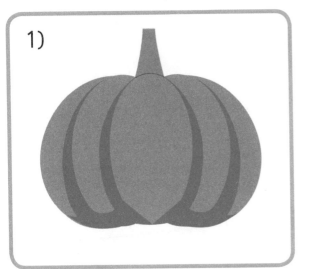

2)

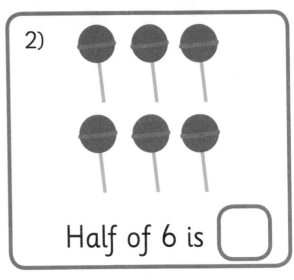

Half of 6 is ▢

3)

Half of 8 is ▢

4)

Teacher's notes

1) Draw a line on the pumpkin to share it fairly into 2 parts.

2) Draw a ring around half of the lollipops and complete the sum.

3) Draw a ring around half of the strawberries and complete the sum.

4) Colour the shapes that show halves.

Fair sharing

Objective

I can share out a group of items equally into smaller groups.

1)

2)

3)

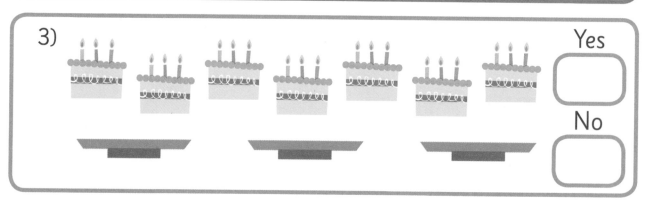

Yes

No

Teacher's notes

1) Share 12 balloons fairly between 4 children.

2) If each child gets 5 sweets, how many children are there?

3) Can 7 cakes be shared fairly between 3 children?

Show me the money

Objective

I can identify all coins to £2.

I can apply addition and subtraction skills and use 1p, 2p, 5p and 10p coins to pay the exact value of items to 10p.

1)

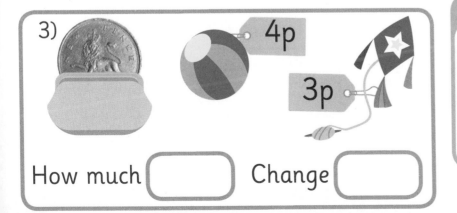

2)

8p   6p   9p

3)

4p   3p

How much [ ]   Change [ ]

Objective

I can link daily routines and personal events to time sequences.

I can name the days of the week and know the months of the year in sequence.

I can name features of the four seasons.

1)

2)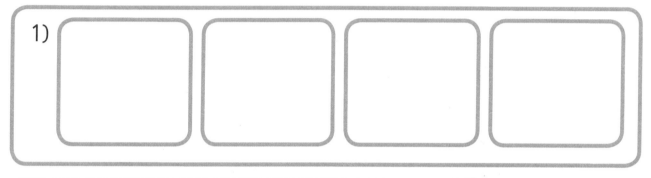

Teacher's notes

1) Draw a picture for each season: Spring, Summer, Autumn, Winter

2) Discuss, order and number the pictures: 1, 2 and 3.

3) Copy the days of the week into the correct place so that the days of the week are in order.

3)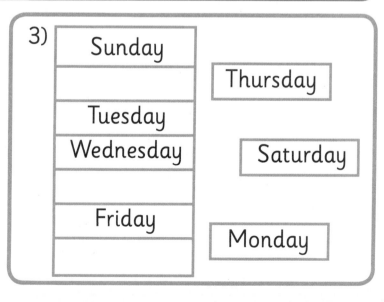

Sunday

Thursday

Tuesday

Wednesday

Saturday

Friday

Monday

Date completed

Objective

I can use timing devices.

I can read and show o'clock times in analogue and digital forms.

I can use appropriate language when discussing time, including before, after, o'clock, hour hand, minute hand.

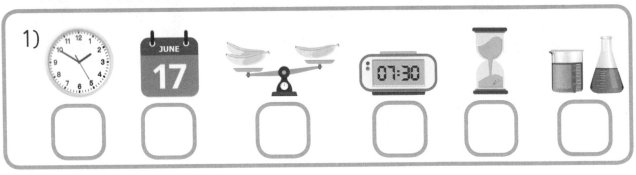

1)

2)

3) o'clock   :   o'clock   :

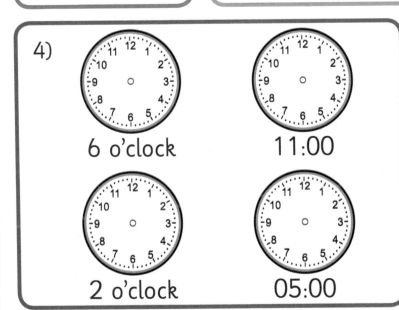

4)

6 o'clock   11:00

2 o'clock   05:00

Teacher's notes

1) Tick the pictures that show time measurers.

2) Complete the clock to show 5 o'clock.

3) Write the correct time under each clock.

4) Read the time and draw the hands on the clock.

Objective

I can describe common objects using appropriate measurement language.
I can compare and describe lengths, heights, mass and capacity using language.

1)

a   b   c

2)

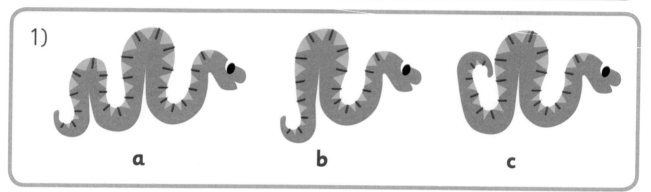

3)

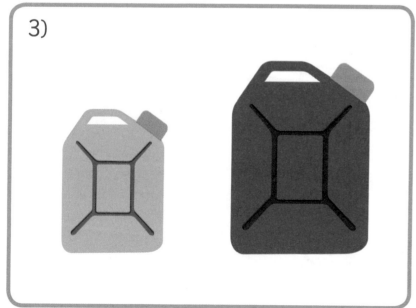

Teacher's notes

1) Circle the longest snake.

2) Tick the heaviest animal toy.

3) Circle the water canister that will hold most.

How many?

Objective

I can share relevant experiences in which measure is used.

I can estimate then measure the length, height, weight, mass and capacity using a range of non-standard units.

I can demonstrate the skills of estimation in the context of number and measure using relevant vocabulary.

1)

Guess: [ ]   Measure: [ ]

2)

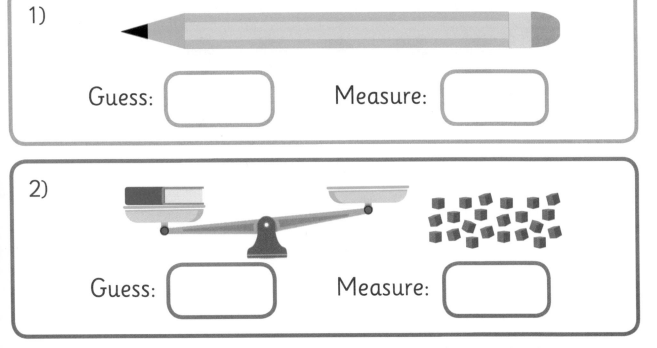

Guess: [ ]   Measure: [ ]

3)

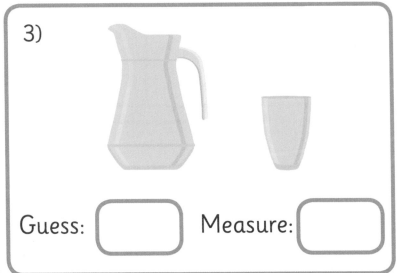

Guess: [ ]   Measure: [ ]

Teacher's notes

1) Guess how many cubes long this pencil is. Measure the pencil using cubes.

2) Guess how many cubes heavy this book is. Measure how many cubes heavy this book is. You will need a pan balance, cubes and a book to weigh.

3) Guess how many cups will fill the jug. Measure how many cups fill the jug. You will need a cup, a jug and dried rice, lentils or water.

Dancing patterns

Objective

I can copy, continue and create simple patterns.

1)

2)

| 1 | 2 | 1 | 2 | | | | |

3)

4)

Teacher's notes

1) Copy the pattern that you see on the scarf in the picture, onto the blank scarf.
2) The bears are marching. Continue the number pattern that they are marching to.
3) Draw an AB pattern using a circle and a rectangle.
4) Draw an ABC pattern using shapes.

24

Carry on please

Objective

I can explore, recognise and continue simple number patterns.

1)

| 20 | | 18 | 17 | | 15 | 14 | | 12 | | 10 | 9 | | | 6 | 5 | | 3 | | 1 |

2)

| ||| | ⅲ̷ | ⅲ̷ | || | ||| | ⅲ̷ | ⅲ̷ | || | | | | |

3   5   5   2   3   5   5   2   ◯◯◯◯

3)

Balloons: 13 8 19 4 18 12 15

1 2 3 ◯ 5 6 7 ◯ 9 10

11 ◯ ◯ 14 ◯ 16 17 ◯ ◯ 20

Teacher's notes

1) Write in the missing numbers.
2) Draw sticks to continue the pattern. Write your number pattern below.
3) Fill in the missing numbers on the number line using the balloons.

Date completed

Objective

I can recognise, describe and sort common 2D shapes and 3D objects.

1)

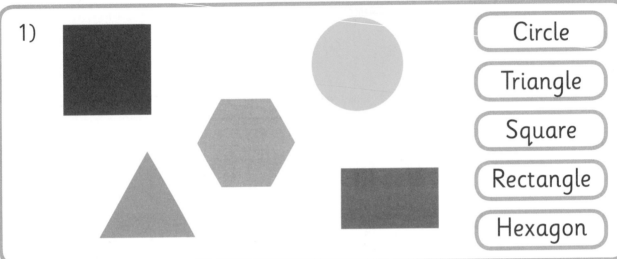

- Circle
- Triangle
- Square
- Rectangle
- Hexagon

2)

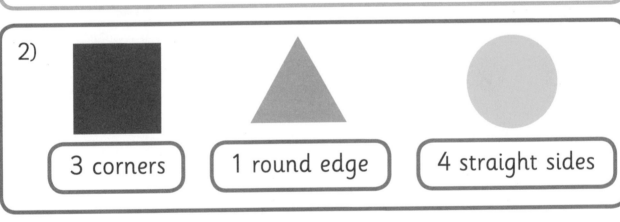

3 corners | 1 round edge | 4 straight sides

3)

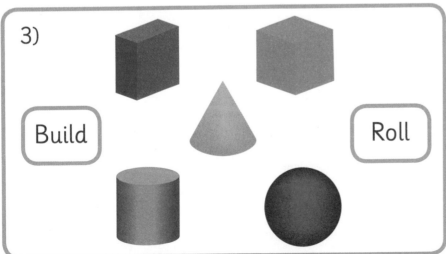

Build   Roll

Teacher's notes

1) Match the shapes to the correct names.

2) Match the shape to its description.

3) Which shapes are good to build with and which will roll? Match the shape to the correct box.

Date completed

Objective

I can talk about position and direction.

1)

2)

1	2	3	4	5	6	7	8	9	10
11	12	13	14	15	16	17	18	19	20
21	22	23	24	25	26	27	28	29	30

Teacher's notes

1) Draw a car in **front** of the bus. Draw a lorry **behind** the bus.

2) Circle the number **to the left of** 22. Circle the number **above** 15.
   Circle the number **below** 20. Circle the number **to the right of** 11.

Symmetry

Objective

I can identify, describe and create symmetrical pictures.

1)

2)

Teacher's notes

1) Tick the objects that are symmetrical.

2) Design a symmetrical butterfly. Tell me about your design.

Sorting toys

Objective

I can collect and organise and sort objects.

You need

Coloured pencils

1)

Wheels

Teddy bears

Arms

2)

Teacher's notes

1) Tell me how you would sort these toys between Finlay and Nuria?
   How many toys have wheels?
   How many toys are teddy bears?
   How many toys have arms?

2) Colour the fruit to sort it.
   Tell me another way of
   sorting the fruit.

Objective

Objective

I can use and make displays of data.

1)

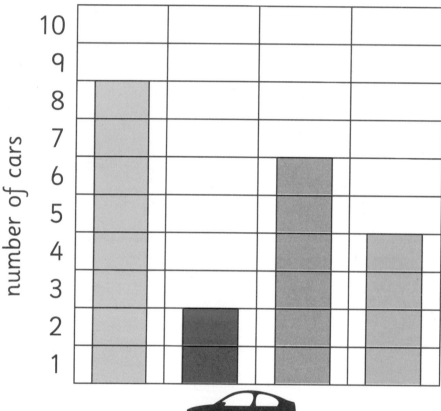

Colour of cars passing school

number of cars

colour of cars

Teacher's notes

This diagram shows the colours of cars passing the school at breaktime.

1) How many yellow cars?

2) How many red cars?

3) How many more blue cars than green cars?

4) How many red cars and green cars altogether?

5) Which colour car was seen the most?

6) Which colour car was seen the least?

1)

Number of shapes in the picture

number of shapes

type of shape

2)

Teacher's notes

1) Colour the blocks to create a block graph to show how many of each shape.

2) Tell me what each of these signs are. Where might I see them? How would I use them to plan a journey?

Primary Maths for Scotland
Early Level Record Book

Authors: Julie Brewer, Sheena Dunlop and Lesley Ferguson
Series Editor: Craig Lowther

© 2020 Leckie

001/05022020

10 9 8 7

The authors assert their moral right to be identified as the authors for this work.

ISBN 9780008359706

Published by
Leckie
An imprint of HarperCollinsPublishers
Westerhill Road, Bishopbriggs, Glasgow, G64 2QT
T: 0844 576 8126   F: 0844 576 8131
leckiescotland@harpercollins.co.uk   www.leckiescotland.co.uk

HarperCollins Publishers
Macken House, 39/40 Mayor Street Upper, Dublin 1, D01 C9W8, Ireland

Publisher: Fiona McGlade
Managing editor: Craig Balfour
Proofreaders: Lauren Reid, Louise Robb

A CIP Catalogue record for this book is available from the British Library.

Acknowledgements
Images © Shutterstock.com

Whilst every effort has been made to trace the copyright holders, in cases where this has been unsuccessful, or if any have inadvertently been overlooked, the Publishers would gladly receive any information enabling them to rectify any error or omission at the first opportunity.

Printed in the UK by Martins the Printers.

Also available for Early Level:

Teacher Guide
978-0-00-835969-0

Digital Pack
978-0-00-835968-3

Wipe-clean Templates
978-0-00-836445-8

ISBN 978-0-00-835970-6

9 780008 359706